Jarrold Nature Series

Photographs and text by

Heather Angel, MSc, FIIP, FRPS

British Wild Orchids

Jarrold Colour Publications, Norwich

1 ×0·8

1. EARLY-PURPLE ORCHIDS (*Orchis mascula*). This is both one of our most wide-spread orchids and also one of the earliest to flower from early April onwards. Here the plants are seen growing alongside cowslips on chalk grassland. Early-purple orchids also grow on roadside verges and in coppiced woodlands (**7**), typically on calcareous or sometimes on neutral soils. The pleasant smell of the opened flowers becomes distinctly cat- or goat-like after fertilisation has taken place. Shakespeare describes them as 'long purples' – one of Ophelia's garland flowers – in *Hamlet*.

Wild orchids have for long attracted the attention of botanists and naturalists alike. The exquisite beauty of the flowers and the sporadic appearance of several species have both contributed to the thrill and excitement involved in searching for orchids in the wild. While most readers will be familiar with some local orchid sites in their neighbourhood, it is hoped that this book will encourage the search of other areas so that new sites may come to light. If you should be lucky enough to find one, tell your local Trust for Nature Conservation.

June is *the* month for orchids, since more species flower during this month than at any other time. It is often difficult to give precise flowering times, because they will vary depending on the weather. Flowering times are also later in the north of Britain than in the south, and the aspect of a site is an important factor; orchids growing on south-facing slopes flower earlier than those on north-facing slopes. Although orchids can be found growing in a wide variety of habitats, including beech and pine woods, marshes, bogs and fens as well as heathlands and sand dunes, the majority grow on limestone and other calcareous soils. Chalk grassland areas in southern England are comparatively rich in orchids, partly because the orchids prefer calcareous well-drained sunny localities, but also because this area is the northern boundary of their recolonisation of Britain after the last glaciation. Unlike a beechwood or an oak wood, both of which are known as climax communities, chalk grassland represents an intermediate stage in natural plant succession. Grazing by sheep, cattle or rabbits helps to keep in check the progression towards a natural beech forest climax. Apart from some steep dry hillsides with little soil cover, ungrazed chalk grassland will gradually become invaded by tall grasses and shrubs which soon choke out the orchids. The Nature Conservancy Council rent grazing to local farmers on chalk grassland reserves to help maintain the short turf which is so essential for many of the orchids. They do have a remarkable ability to persist in areas where they once abounded, but which have subsequently become choked with other vegetation. One piece of woodland in Hampshire which was cleared of thick undergrowth, produced great numbers of helleborines over the following few years. Scrub clearance on downland will also revive an orchid site so long as the scrub has not persisted for too long.

Orchid seeds are so minute that they can be carried for considerable distances in air currents. Some of our species may even be reinforced, in the south-east in particular, by seed being blown across the Channel from the Continent. Even when seed lands on suitable soil, growth of the seedlings is so slow that leaves may not be produced for two or three years. The time taken for an orchid to flower after germination is very variable; but it is always several years and, in the case of the common twayblade and the dwarf or burnt orchid, it is thirteen to fifteen years. An interesting and important part of the orchid's development is the close association – known as a mycorrhiza – which the root forms with a fungus. The fungus obtains its food from the plant remains – the humus – in the soil and passes some of this to the young orchid before it develops the green leaves. The fungus continues to feed the brownish bird's-nest orchid throughout life.

The rarity of many of our wild orchids, coupled with the many years of development, make it imperative that no orchid – however common it may appear in a small area – should be picked.

2 ×0·3

2. BIRD'S-NEST ORCHIDS (*Neottia nidus-avis*) with **WHITE HELLEBORINES** (*Cephalanthera damasonium*). Beechwoods are the typical habitats of these orchids. The bird's-nest orchid is brownish yellow all over. It has no leaves and since it contains no green chlorophyll it cannot manufacture its food. Instead, it feeds on nutrients in the leaf litter humus. These are obtained via a fungus which lives inside the orchid roots. The shade-loving bird's-nest orchid should not be confused with the yellowish-brown broomrapes which also have no chlorophyll, but which grow in more open situations. Broomrapes live as parasites on green plants. The short fleshy roots of the bird's-nest orchid resemble an untidy bird's nest.

3. BIRD'S-NEST ORCHID (*Neottia nidus-avis*). After germinating, it takes eight years before a flowering spike is produced. Even when quite fresh, from a distance the brownish flowers look more dead than alive. Although this orchid is found throughout Britain, it is more common in the south.

4. LARGE WHITE HELLEBORINE

(*Cephalanthera damasonium*). Often found in beechwoods, this orchid is confined to chalky soils in the south and south-east of Britain. The erect green parallel-veined leaves can be seen in Plate 2. Creamy egg-shaped flowers, each with a lower yellow lip, give rise to the alternative names of egg orchid and poached-egg plant. There are usually 3–8 flowers, which rarely open any more widely than are shown in this photograph, and can pollinate themselves. They tend to grow in towards the stem or only slightly curled away from it. The striking green-stemmed and white-flowered plants stand out well against a carpet of fallen beech leaves. Reaching up to 45 cm high, they often dwarf neighbouring bird's-nest orchids (**2**). Both these orchids flower from mid-May to early June.

5. NARROW-LEAVED HELLE-BORINE (*Cephalanthera longifolia*).

This orchid also favours chalky soils, especially shady areas. Although it has a widespread distribution over Britain, it is by no means an abundant species. All the three photographs on this spread were taken on the same day at the end of May in a southern beechwood. Notice the much longer and narrower leaves and the more open and outwardly spreading flowers of this species. The white flowers are cross-pollinated by small bees.

6. NARROW-LEAVED HELLE-BORINE (*Cephalanthera longifolia*).

Also known as the sword-leaved helleborine, this close-up shows the long green ovary at the base of each flower. Apart from the lowermost flowers, the bract at the base of each ovary is much smaller; whereas the large white helleborine has bracts which are longer than the ovaries.

4 ×1

5 ×0·25

6 ×1

7

8 ×1

7. EARLY-PURPLE ORCHIDS (*Orchis mascula*) with **OXLIPS** (*Primula elatior*). These orchids, already illustrated in Plate 1, are seen here growing in a clearing of a coppiced Essex woodland. They will not grow in very shady woods. The spotted leaves show best on the extreme left-hand plant. Even though these orchids are abundant, their numbers in any one particular location can fluctuate very widely from one year to another. In past ages throughout Europe, orchids – and this one in particular – were taken as an aphrodisiac. The word *Orchis* means testicle and the root tubers were clearly symbolic.

8. BROAD-LEAVED HELLEBORINE (*Epipactis helleborine*). This species is unlikely to be overlooked, since it grows up to 100 cm high. It often grows in the shade of woods, especially beechwoods in the southeast. Although it prefers chalky soils, it will also grow on acid soils. The flowers – which are very variable in number and colour – open in July to August. At least some parts of the flower usually appear pinkish in colour. This particular orchid was growing in the same beechwood as the helleborines on the previous spread; although at the time they were photographed, this one had developed only its leaves.

9. WASP POLLINATING BROAD-LEAVED HELLEBORINE (*Epipactis helleborine*). The nectar produced by these flowers is especially attractive to wasps, which cross-pollinate the flowers as they fly or crawl from one to another. Several of the cream-coloured pollinia can be seen attached to the head of this wasp, which spent over three hours feeding at a single flower spike.

10. CORAL-ROOT ORCHID

(*Corallorhiza trifida*). These small yellowish-brown orchids are confined to the north of England and Scotland where their distribution is very local. They grow, as shown here, in pinewoods and also beneath birches or among marram grass in sand dunes. Coral-root has no true roots, only a branched coral-like creamy rhizome. Like the bird's-nest orchid, this rhizome contains a fungus from which the orchid obtains its food. Both the flower stalk and the ovaries of coral-root appear slightly greenish, but otherwise it contains no chlorophyll and has no leaves. The flower spikes appear when enough food reserves have accumulated in the rhizome, and the flowers open in June to July.

11. LESSER TWAYBLADE (*Listera cordata*).

This is another northern species, which grows also in North Wales and parts of the south-west. Although it is much smaller than the more widespread common twayblade (**25**), it also has a pair of leaves clasping the lower part of the stem, but these are heartshaped. Lesser twayblade grows on acid soil in pinewoods as well as on moorlands and in bogs. When growing among heather, the orchids often cannot be seen unless the heather plants are carefully separated. The flowers appear from mid-June to mid-August.

12. HEATH SPOTTED ORCHID

(*Dactylorhiza maculata ericetorum*). This orchid also grows in acid locations, notably heathlands, but also bogs. If the shape of an individual flower is compared with a spotted orchid flower (**19**), the dashed markings are more numerous and the side lobes of the lip are broader. The paired pollonia can be seen inside several flowers.

11 ×1
12 ×1·5

13 ×1·5
14 ×0·2

13. BEE ORCHID (*Ophrys apifera*).

This striking orchid is one of many species which grow on chalk grassland areas, especially in the south of England. It does not grow in Scotland. Unlike the fragrant orchid, the bee orchid now rarely occurs in large numbers. However, the tall spike, which can grow up to 40–50 cm high, and have two to seven widely spaced flowers on it, is not easily overlooked in June. Each flower has three pinkish sepals. It is the velvety-brown lip which gives the flower its resemblance to a bee. In the Mediterranean region this resemblance lures bees which attempt to mate with the flowers, and in so doing, may cross-pollinate them. In Britain, however, insects rarely visit bee orchids which are self-pollinated. Bee orchids grow also in disused gravel pits, as well as woods.

14. FRAGRANT OR SCENTED ORCHID (*Gymnadenia conopsea*).

Also growing on chalk grassland, the numbers of this orchid can fluctuate widely from one year to another, but the tall pinkish spikes always stand out well against the grassy habitat. The flower colour is very variable, as can be seen from the two plants illustrated on this spread. Occasionally forms with white flowers appear. The orchid can, however, be recognised by the unspotted flowers with their long spurs (**15**), which open in June.

15. FRAGRANT ORCHID (*Gymnadenia conopsea*).

The long spurs are filled with nectar, which can be reached only by long-tongued insects such as butterflies and moths. These are the insect-pollinators of this orchid. The powerful sweet scent lasts throughout the night. After the flowers have been fertilised, it becomes decidedly rancid.

15 ×

16. GREEN-WINGED ORCHID (*Orchis morio*). Although this orchid could perhaps be mistaken for the early-purple orchid (**1**) it has no spots on the leaves and the purple sepals are marked with obvious green veins. Also, it grows out in the open and not in woodlands. Usually growing on chalk but also on clay, the flowers open in May or early June. They are vanilla scented and pollinated by social and solitary bees. Local names for this orchid include Bloody Man's Finger (Cheshire) and Goosey Gander (Dorset and Somerset).

17. LESSER BUTTERFLY-ORCHID (*Platanthera bifolia*). Growing on both chalky and acid soils, this orchid can be confused with the greater butterfly-orchid. The latter is often a larger plant, but the best distinguishing feature is the position of the pollen sacs inside the flowers. In the lesser butterfly they are parallel, in the greater they are arranged far apart in the form of an inverted V. Like the fragrant orchid, the flowers have long nectar-filled spurs, which attract butterflies and moths, especially night-flying moths. As the moth feeds, the pollinia become attached to its tongue and bend forward ready to pollinate the next flower the insect visits.

18. BURNT ORCHID (*Orchis ustulata*). Also known as the burnt-tip or dwarf orchid, it is a local species of chalky soils. It is the dark-coloured flower buds which give the burnt-looking appearance. The almond-scented flowers are pollinated by insects. One of the smallest grassland orchids, it will not survive growing among tall grasses. It may be as long as thirteen or fourteen years after germination, before a flower spike is produced.

19. COMMON SPOTTED ORCHID (*Dactylorhiza fuchsii*). This is a widespread and varied orchid, which occurs throughout Britain. Since it often grows out in the open, on grassy slopes and roadside verges, it is not easily overlooked. Even when growing in open woodlands, the tall spikes stand out among the other plants. The leaves are often spotted, but sometimes there may be no spots at all. The tightly packed pinkish flowers marked with darker lines, open during June to July. They are pollinated by insects, and large numbers of seeds are produced. The Somerset name of Dead Man's Finger refers to the flattened undergound tubers which resemble fingers on a hand. Another Somerset name is Adam and Eve.

20. MAN ORCHID (*Aceras anthropophorum*). Even though they grow out in the open, man orchids can be overlooked because their yellowish-green flowers blend in well with the grasses and other plants of chalk downland. The plants often grow near the bottom of slopes. The flowers have an unpleasant smell which attracts hoverflies and midges.

21. MAN ORCHID (*Aceras anthropophorum*). When each flower opens fully during May or June, the resemblance to a little 'man' can be seen. The 'head' is formed from the sepals and petals and the 'arms' and 'legs' from the lip. Sometimes the edges of the 'head' are marked with red. Man orchids grow only in the south and south-east of England. The height of the plants varies from 10 to 50 cm. If the surrounding vegetation is short, the orchids will likewise remain as short spikes, but among high grasses, the orchids will grow up higher.

20 ×0·4
21 ×1·2

22. PYRAMIDAL ORCHID (*Anacamptis pyramidalis*). It is when the lowermost flowers open, that the distinctly pyramidal shape is best seen. By the time all the flowers have opened, the pointed top has become much more rounded. Pyramidal orchids can often be found in southern England growing alongside fragrant orchids on chalk grassland. They also grow in open scrub and sand dunes. Both the leaves and the flowers are unspotted. Pyramidal orchids tend to be bright pink in colour; whereas fragrant orchids have a suggestion of a lilac tint. In the centre of the lip is a pair of ridges which lead into the mouth of the long spur.

23. PYRAMIDAL ORCHIDS (*Anacamptis pyramidalis*). These orchids were flowering on a sand dune in South Wales in June. The strong sweet smell of the flowers attracts butterflies and moths, especially burnet moths. The proboscis or tongue of the insect is guided into the spur by the lip ridges seen in Plate **22**. The ripe orchid pollinia then stick to the proboscis, bending forward ready to pollinate the stigmas of another flower. Sometimes several pairs of black pollinia can be seen on the proboscis of a six-spot burnet moth.

24. FROG ORCHID (*Coeloglossum viride*). Growing all over Britain on open, well-drained calcareous soils, the small greenish flowers make it an inconspicuous plant which is easily overlooked. It will grow at high altitudes; this particular plant was found at 650 m. The origin of the name 'frog' is curious, since it bears no resemblance to the animal. The flowers usually open in July. Their slight honeysuckle scent attracts bees and beetles.

23 ×1

24 ×1

25. COMMON TWAYBLADE (*Listera ovata*). The distinctive paired leaves – known locally as 'sweethearts' – enable this orchid to be recognised even before the flower spike appears. One of the commonest of our orchids, twayblade grows both on grassy slopes and also in woodlands. It is also extremely slow-growing, taking up to twelve years before it flowers. The inconspicuous green flowers open from May to June. The broad lip has a distinct notch at the lower end. Leading up from the notch is a nectar-filled groove, which attracts small insects, including ichneumon wasps. These wasps lay their eggs on other insects and their grubs live as parasites inside them.

26. FLY ORCHID (*Ophrys insectifera*). Related to the bee orchid (**13**) the fly is even more convincing in its resemblance to an insect. When growing in the shade of woodlands, the tall stems with their widely spaced small reddish-brown flowers are difficult to spot. The fly orchid, although widespread over a large part of Britain, is local and it flowers in May or June.

27. FLY ORCHID (*Ophrys insectifera*), **DETAIL.** Each flower has three pale-green sepals and a reddish-brown lip with a blue band across the middle. The pair of narrow brown petals which project upwards, resemble the 'antennae' of the 'fly'. The close resemblance of the flowers to an insect with the additional attraction of a scent, lures small wasps to the flowers. The male wasps attempt to mate with the flower mimics and in so doing transfer the pollinia from one flower to another. This behaviour – known as pseudo-copulation – has been observed on European fly orchids but not, so far, in Britain.

28. LIZARD ORCHID (*Himantoglossum hircinum*). The bizzare-shaped flowers, usually on a tall spike, ensure that this striking orchid is not easily overlooked in July. The long twisted lip resembles the tail of a 'lizard', while the flowers are said to smell of goats. They are visited by flies and other insects. With the exception of a few established colonies, this orchid has a spasmodic appearance, mainly in southern and eastern England. Therefore, when this orchid does appear, it often hits the headlines. Lizard orchids grow on chalky soil and also on older sand dunes.

29. AUTUMN LADY'S-TRESSES (*Spiranthes spiralis*). This is one of the latest flowering of our wild orchids. The spiral arrangement of the small white flowers resembles a plait of hair. Opening in September, they emit a jonquil-like scent during the day, which attracts bees. As shown here, clusters of these orchids can be found on sunny slopes of chalk downs and verges where the grass is short. The distribution of autumn lady's-tresses lies south of a line running across from the Lake District.

30. JERSEY OR LOOSE-FLOWERED ORCHID (*Orchis laxiflora*).

As the name suggests, this orchid is known in Britain only from the Channel Islands, where it grows in damp meadows and marshy areas. At first sight, it resembles an early-purple orchid (**1**), but on closer examination, the flowers are much more widely spaced on the stem. Also, the leaves have no spots, and the lip is divided into two, not three, lobes. The spur, which may be horizontal or turned upwards as shown here, has a notch at the tip. The reddish bracts are almost as long as the ovary. The strong colour as well as the tall spike both help to separate these orchids from among the neighbouring tall grasses and other plants. May to June is the time when this orchid flowers.

31. MUSK ORCHIDS (*Herminium monorchis*).

Like the autumn lady's-tresses (**29**) these orchids can be found growing among short turf on chalk grassland, where they can withstand periods of drought. Their distribution is much more restricted, however, being confined to the south and south-east part of England. The plants are small, often only 7–12 cm high, but sometimes reaching 20 cm. Since most of the plants are produced vegetatively from underground tubers, they tend to occur in neat clusters. Musk orchids usually flower during June. The tiny yellowish-green somewhat bell-shaped flowers produce a heavy honey-like scent rather than musk. They are pollinated by small insects, some of which crawl over the flowers, while others, such as chalcid wasps, fly from flower to flower. The musk orchid, which is one of the smallest British orchids, is difficult to spot casually in the field, on account of its size and greenish flowers.

32. MARSH HELLEBORINE (*Epipactis palustris*).

Unlike the other helleborines illustrated in this book, the marsh helleborine prefers to grow out in the open instead of in shady places. This orchid is most likely to be seen growing in marshy areas which are fed with alkaline water. It does not grow in acid boglands. It also grows – often in large numbers – in the slacks of older sand dunes. In this situation, the plants are much smaller than when they have to grow up among lush marshland vegetation. When the flowers open in July to August, they appear white. Seen close up, the three outer segments have brown markings, while the inner two have distinct purple veins. Nectar is contained in a cup at the base of the lip. The outer part of the lip, which is white and frilly with a central yellow spot, is connected to the cup by a slender elastic 'waist'. When an insect – usually a bee – alights on the lip, it bends the outer part downwards. Then when it leaves the flower, the lip springs upwards making the bee brush against the pollen sacs in the upper part of the flower.

33. NORTHERN MARSH OR DWARF PURPLE ORCHIDS (*Dactylorhiza purpurella*).

The dark-purple flower spikes are easily seen in marshy areas in the north and west parts of Britain, from the end of June through July. This is the latest of the marsh-orchids to come into flower.

34. EARLY MARSH-ORCHIDS (*Dactylorhiza incarnata*).

These marsh-orchids which flower much earlier in late May through June, also grow in water meadows and marshes. The sides of the lip fold downwards so that it appears very narrow.

35. MARSH FRAGRANT OR-CHID (*Gymnadenia conopsea densiflora*). This variety of the fragrant orchid has a much denser flower spike. It usually grows in much wetter places than the fragrant orchid (**15**). The flowers open several weeks later and have a pleasant carnation scent. This scent is common to several flowers which are pollinated by moths and butterflies.

36. BOG ORCHID (*Hammarbya paludosa*). Our smallest native orchid grows in wet acid bogs. The inconspicuous yellow-green flowers, coupled with the small 3–12 cm high plants, make it a difficult orchid to spot against the sphagnum moss cushions. It has a scattered, local distribution throughout the country. The flowers are unusual in that the lip is at the back or top instead of the bottom. This position is due to the ovary twisting through a complete 360° circle. The flowers, which open in July and August, are visited by insects and a fly can be seen on the upper left side of the spike. Bog-orchid seeds float on water and so can be carried in bog water some distance from the parent plant. The roots are tiny hair-like structures which contain a fungus.

37. FEN ORCHID (*Liparis loeselii ovata*). Also frequenting wet places, is the rarer fen orchid; which has a much more limited distribution than the bog orchid and it grows only where the pH of the water is alkaline or neutral. This variety, the broad-leaved fen orchid, grows in the dune-slack systems of South Wales and North Devon. It has broader leaves than the typical subspecies which is confined to East Anglia. Here it is growing among creeping-willow plants. The fen orchid flowers in June to July.

38. MONKEY ORCHID (*Orchis simia*).

This orchid gets its name from the resemblance of each flower to a small monkey. Like the man orchid (**21**), the hood forms the 'head', while the five-lobed lip forms the 'arms', the 'legs' and the 'tail'. The monkey flower is unique among British orchids, in that the flowers at the top of the spike are the first to open on most plants. No nectar is produced, but visiting insects include bees which may be attracted by a sugar secretion. This orchid, like the military orchid on the facing page, is now very rare in this country, but both orchids are much more widespread in Europe. Several factors have contributed to the decline of the monkey orchid in Britain; the most obvious one being agricultural development of the grassland sites. Additional factors include land lost to building and in the past, the collection of plants. Monkey orchids come into flower in the last part of May. They smell faintly of coumarin, which is a fragrant but volatile chemical substance produced by newly cut hay.

39. MILITARY OR SOLDIER ORCHID (*Orchis militaris*).

This rare orchid was thought to be extinct in Britain until 1947 when a colony was discovered some twenty-five years after the last known record. This demonstrates the possibility of new orchid sites coming to light in suitable locations. Like the monkey orchid, it flowers in late May/early June and grows on calcareous soils. It grows on the edges or in open parts of woods rather than on open grassland sites. The 'legs' of each flower are much wider than on the monkey, and there is no obvious 'tail'. This orchid prefers to grow in places in which the climate has warm, rather than hot, dry summers.

39 ×1

40. LADY ORCHID (*Orchis purpurea*). The tall plant (20–90 cm) and the striking flowers make this one of the most handsome of all our wild orchids. Although both its size and its preference for scrub and woodland sites cannot cause confusion with the burnt orchid (**18**), both orchids have flowers with dark-coloured hoods and pale-pink spotted lips. The lady orchid flowers, like many woodland plants, early in the season. It is a magnificent sight to see dozens of lady orchids flowering in a woodland setting in early May.